LET'S LOOK AT LANDFORMS

THOSE ARE PLAINS!

BY DWAYNE HICKS

Gareth Stevens PUBLISHING

Please visit our website, www.garethstevens.com. For a free color catalog of all our high-quality books, call toll free 1-800-542-2595 or fax 1-877-542-2596.

Library of Congress Cataloging-in-Publication Data
Names: Hicks, Dwayne, author.
Title: Those are plains! / Dwayne Hicks.
Description: New York : Gareth Stevens, [2022] | Series: Let's look at landforms | Includes index.
Identifiers: LCCN 2020032280 (print) | LCCN 2020032281 (ebook) | ISBN 9781538263877 (library binding) | ISBN 9781538263853 (paperback) | ISBN 9781538263860 (set) | ISBN 9781538263884 (ebook)
Subjects: LCSH: Plains–Juvenile literature. | Great Plains–History–Juvenile literature.
Classification: LCC GB572 .H52 2022 (print) | LCC GB572 (ebook) | DDC 551.45/3–dc23
LC record available at https://lccn.loc.gov/2020032280
LC ebook record available at https://lccn.loc.gov/2020032281

Published in 2022 by
Gareth Stevens Publishing
29 E. 21st Street
New York, NY 10010

Designer: Tanya Dellaccio
Editor: Greg Roza

Photo credits: Cover Westend61/Getty Images; pp. 3–22 (background texture) MicroOne/Shutterstock.com; p. 5 Brent_1/iStock/Getty Images Plus/Getty Images; p. 7 Iñigo Fdz de Pinedo/Moment/Getty Images; p. 9 Alexander Kolbasov/TASS/Getty Images; p. 11 Kenneth Canning/E+/Gety Images; p. 13 Jeff Greenberg/Universal Images Group/Getty Images; p. 15 Lana2011/iStock/Getty Images Plus/Getty Images; p. 17 beppeverge/Moment/Getty Images; p. 19 Cavan Images/Cavan/Getty Images; p. 21 Carl Young/EyeEm/Getty Images.

Printed in the United States of America

CPSIA compliance information: Batch #CSGS22: For further information contact Gareth Stevens, New York, New York at 1-800-542-2595.

CONTENTS

Boldface words appear in the glossary.

Flat Land

Plains are **broad**, flat areas of land. They're often covered with grass and trees. Plains are so big they're often used for farming. Plains form at lower **elevations**. Plateaus are flat areas of land that form at higher elevations.

Lava Plains

Earth's first plains were most likely formed from **lava**. Lava sometimes pours out of cracks in Earth's **surface**. The lava flows out over the land and makes new land. When the lava covers a large area, a plain can form.

Made by Erosion

Some plains are formed by **erosion**. Wind, water, and ice wear away mountains and hills. Pieces of rock and dirt come to rest in lower areas. Over many years, **layers** of rock pile up. This creates a plain.

Grasslands

Grasslands are plains on which grass and other plants grow. Many animals live and **graze** on grasslands.

There are many different names for grassland plains. They may be called plains or prairies. Drier grasslands are called savannas.

Coastal Plains

Coastal plains form where some land meets the ocean. These plains rise gently as one moves from the ocean to land. They may end at a mountain or other landform. The Florida Everglades is on a coastal plain.

Floodplains

Some rivers **flood** the land around them. This often happens in spring when rivers are full. The flat area that surrounds a river is called the floodplain. This is the area that gets covered with water during a flood.

Alluvial Plains

Floodplains can grow, move, and join together. This creates a larger plain called an alluvial plain. "Alluvial" means something is made up of the small pieces of rock created by erosion. Rivers carry those pieces until they come to rest.

Plains and People

Plains have long been important to people. Over one-third of Earth's surface is covered with plains. The first towns formed along plains and rivers. These places were good for travel and farming. The first large farms were planted on plains.

The Great Plains

The Great Plains are a huge plain in North America. It covers 1,125,000 square miles (2,900,000 sq km) of land! It's home to many towns, cities, and farms. It's also home to many kinds of wild plants and animals.

GLOSSARY

broad: reaching far and wide

elevation: the height something is above sea level

erosion: the wearing away of the earth by wind or water

flood: a large amount of water covering an area of land that is usually dry

graze: to eat grass in a field

lava: melted rock that comes to Earth's surface

layer: one part of something lying under or over another

surface: the outer layer of something

FOR MORE INFORMATION

BOOKS

Capelle, Jenna. *Plains.* Mendota Heights, MN: North Star Editions, 2018.

Somervill, Barbara A. *Plains and Plateaus.* North Mankato, MN: Cherry Lake Publishing, 2014.

WEBSITES

Plain

www.nationalgeographic.org/encyclopedia/plain/

Read about what plains are and see photos on National Geographic's website.

Plain Facts for Kids

kids.kiddle.co/Plain

Want to learn more about plains? This website has more information about specific plains around the world!

INDEX